Terraforming
by Patrick H. Stakem

(c) 2018

Number 27 in the Space Series

Table of Contents

Introduction..3
Author..3
Terra-forming..4
The planets of our solar system, and some remote outposts. 5
    Mercury..6
    Venus...7
    Earth..10
    Near Earth Objects..11
    The Moon...12
    The Asteroid Belt, dwarf planets, and Centaurs...................14
    Comets...16
    Mars...17
    Gas Giants...22
        Jupiter...22
        Saturn...25
    Ice Giants..28
        Uranus..28
    Neptune..31
    Pluto and beyond...33
Challenges...34
    Power...35
    Radiation Environment..36
Exo-challenges...36
Wrap up...39
Afterword...40
Glossary of Terms..41
Bibliography...46
Resources...55
If you enjoyed this book, you might also be interested in some of these..57

The Earth is just too small and fragile a basket for the human race to keep all its eggs in.
--Robert A. Heinlein

Introduction

This book discusses alternating the environment of a planet to be more Earth-like for human habitation. We will focus on the planets (and moons) of our solar system. There's enough work there to keep us busy for a while, before we consider planets around our neighbor stars.

Author

The author has a BSEE in Electrical Engineering from Carnegie-Mellon University, and Masters Degrees in Applied Physics and Computer Science from the Johns Hopkins University. During a career as a NASA support contractor from 1971 to 2013, he worked at all of the NASA Centers. He served as a mentor for the NASA/GSFC Summer Robotics Engineering Boot Camp at GSFC for 2 years. He taught Embedded Systems for the Johns Hopkins University, Engineering for Professionals Program, for the Graduate Computer Science Department of Loyola University of Maryland, and for Capitol Institute of Technology. He has done several summer Cubesat Programs at the undergraduate and graduate level.

He began his career in Aerospace with Fairchild Industries on the ATS-6 (Applications Technology

Satellite-6), program, a communication satellite that developed much of the technology for the TDRSS (Tracking and Data Relay Satellite System). At Fairchild, Mr. Stakem made the amazing discovery that computers were put onboard the spacecraft. He quickly made himself the expert on their support. He followed the ATS-6 Program through its operation phase, and worked on other projects at NASA's Goddard Space Flight Center including the Hubble Space Telescope, the International Ultraviolet Explorer (IUE), the Solar Maximum Mission (SMM), some of the Landsat missions, and others. He was posted to NASA's Jet Propulsion Laboratory for the MARS-Jupiter-Saturn (MJS-77), which later became the Voyager mission, which is still operating and returning data from outside the solar system at this writing.

He received NASA's Space Shuttle Program Managers Commendation award, two NASA Group Achievement Awards, and the NASA Apollo-Soyuz Test Program Award. He has completed over 42 NASA Certification Courses. He has led and supported efforts at all of the NASA Centers on terrestrial and planetary missions.

Mr. Stakem has been affiliated with the Whiting School of Engineering of the Johns Hopkins University, Graduate Computer Science Department of Loyola University In Maryland, And Capitol Technology University.

Terra-forming

We don't need to Terra-form another world to live there

– colonies with habitats will work. But in the long run, if we're going to maintain a permanent colony, we will want to make the planet more Earth-like. It will take a while. Decades or centuries. We don't understand the process completely, there will be errors. The term Terra-forming came from a science fiction story in 1942. Carl Sagan suggested the planetary re-engineering of Venus in 1961. He published a Planetary Engineering on Mars article in 1973.

Not everyone agrees that Terra-forming is feasible, practical, or even that it should be done. There are dangers we have not, or cannot anticipate. And we certainly don't want to exterminate existing life. We have a lot of bad examples from colonization on our own planet.

NASA has accepted Planetary Engineering, but likes to call it Planetary Eco-synthesis, or Modeling.

Geo-engineering is Terra-forming of Earth. Let's get this right before we start on vacation locations further out in our solar system.

The planets of our solar system, and some remote outposts.

The International Astronomical Union defined in August 2006 that a planet is a celestial body which is in orbit around a star, has sufficient mass to assume hydro-static equilibrium (i. e. a nearly round shape), and. has cleared the debris from around its orbit. Even the definition of "habitable" is not well defined, and goes according to "life as we know it." Let's look at our neighbors and see what it would take to live there.

Mercury

Mercury, the closest planet to the Sun, is in tidal lock, with one side always facing the Sun. Actually, there is a rare 3:2 spin-orbit resonance, not seen elsewhere. For every two revolutions around the Sun, Mercury rotates three times on its axis. It wobbles a bit, creating a twilight zone that is much less extreme. It has no known moons, or Trojans. Being so close to the Sun, it is difficult to observe the planet and its immediate vicinity. We do know it has a heavily cratered surface. Mercury is a rocky planet, with long narrow ridges observed on the surface.

Mercury has a molten core with abundant iron. Its mantle is made of silicates. The crust is thought to be some 35 km thick.

The surface of the planet resembles Earth's moon, with large plains (mare) and craters. It is thought to be 4.6 billion years old. It has no atmosphere.

Mercury is currently being observed up close by the Messenger spacecraft, and this will increase our knowledge of the planet. We really don't have a way to image or study the sun-facing side. The spacecraft has seen evidence of more than 50 pyroclastic flows from active volcanoes.

The surface temperature of Mercury varies from -170 to 435 degrees Celsius. Water ice has been confirmed in deep craters at the poles. Mercury has no atmosphere per se, due to its low gravity. It does have a magnetosphere, with a field strength about 1% of Earth's.

The eccentricity of Mercury's orbit is 0.2, the most of any planet in our solar system. The axial tilt is almost zero, and the orbit is tilted 7 degrees to the plane of the ecliptic. The orbital eccentricity varies wildly from zero to 0.45 over a period of millions of years.

There is some indication that, since the planet wobbles a bit, there is a permanent, temperate "twilight" zone. The problem is getting there. There's no particular advantage to having humans live there.

Venus

Heavy greenhouse clouds of sulfuric acid trap solar energy, and cause massive global warming on a planetary scale. The surface temperature is high enough to melt some metals. We need to find out what went wrong on Venus, and try to avoid that on Earth. Is it possible to terraform our sister planet?

Venus' atmosphere is 96% carbon dioxide at a surface pressure of nearly 100 times Earth's, a greenhouse gone wild. It has no moons. Venus is roughly Earth-sized, but something went terribly wrong.

The heavily clouded atmosphere makes it difficult to observe Venus. We do know it rotates in the opposite direction to most of the other planets, and we know it has no magnetic field.

Our first approach would be a detailed examination of the planet to ensure here is no indigenous life there. We have a responsibility of respecting other life forms. If not from an ethical standpoint, certainly from a science viewpoint. Alternative forms of life in other environments may have come up with some biological tricks that would be of value to us. We certainly wouldn't want some external group to come to Earth and sterilize the planet, just for their own use. At the temperature of Venus, it is not likely that life forms are present.

Venus is a terrestrial, rocky planet, about the same size as Earth. It has a dense atmosphere of carbon dioxide. It's surface pressure is more than 90 times that of Earth's. Venus is hotter than Mercury in spite of being further from the Sun. It has seen extreme volcanism, but no lava flows have been observed. The surface was shaped by volcanic activity, and Venus has more volcanoes than Earth.

Atmospheric pressures between 48 to 70 km, are similar to Earth's. The main component is carbon dioxide. The thick clouds block a lot of ultraviolet light.

Unfortunately, the rain is sulfuric acid.

The first step is to get rid of the excessive CO_2, as this will help with the temperature problem as well? Know anyone who needs a planet's work of CO_2? Actually, yes, the Martians. On Earth, we have rising carbon dioxide levels which lead to global warming. We need to come up with an approach for the home planet as well, before we turn it into another Venus. Plants are a big help.

The big steps on Venus are to reduce its excessive surface temperature, reduce most of the carbon dioxide/sulfur dioxide atmosphere, and add breathable oxygen. Now, of course, you can never do just one thing. For whatever we do, and how we do it, we need to understand the consequences. Extensive modeling and testing will be required, on a planetary scale. One nice thing to have on the Venetian surface is water. We may need to import that in vast quantities. Nothing is going to happen overnight, and we have to protect what we're doing as it proceeds along slowly. Venus also needs a magnetic field, to cut down on energetic particles from the Sun.

Not much is known about the internal structure of Venus. There have been observations of lightning in the thick cloud layer. It has very high wind speeds. Since it rotates opposite to its orbital direction, there are no seasons. It only has a minor tilt to its spin axis. It has no moon, so no tides. A moon could be handy in producing tides, so perhaps a large asteroid could be towed there. We can set off for Venus every 584 days. Any more or less than that are too costly in terms of fuel. The rules of Celestial Mechanics apply.

Venus orbits the Sun in about 225 (Earth) days, with a small eccentricity. It has no moons, but we could redirect an asteroid for it. Sometime in the future. Due to extreme conditions, we can rule out life on Venus, as we know it, but we better check.

Some consider Venus as potentially more habitable than Mars. This is based on the fact that higher up in the atmosphere, there is a temperate region. Some studies have been made of an aerostat or drifting balloon arrangement that could house scientists. Terra-forming Venus would be a long and difficult process. We would need to drastically alter the atmosphere. In addition, for the long run, we need to make our Terra-forming efforts self-continuing, and self-regulating. Truthfully, we have not done well in this regard with our own planet.

Quite a few smart people have considered the Venus problem, and have written up their results. I think its only a matter of time. One interesting approach is a series of aerostats, drifting high in the atmosphere, where the air pressure is near-Earth normal. That's about 50 km. Breathable air would provide the lift in the Venusian atmosphere. We do have to watch out for wind speeds, which can reach 200 mph.

Why would we commit to such a huge, expensive project? We eventually want more real-estate, and we want a "lifeboat" if something happens to Earth. Like a second house in the mountains if we live along the Hurricane coast. might came in handy.

Earth

Earth is our model. We need to learn a lot more about our planetary dynamics. And how the planet interacts with the Sun. It is our only known planet harboring life.

We observe the Earth continuously with a series of orbiting satellites, to keep track of the weather, and violent events. In fact, Earth is the most observed planet, and we know the most about it. Not that it can't give us surprises. Earth's moon is the largest object in the sky, and has been observed since humans first looked up. It is orbited by a series of satellites, has surface landers, and has been visited by Astronauts from the United States.

The bow shock is plasma from the Sun hitting the Earth's or other planet's magnetosphere. The plasma is ionized, and follows spiral paths along magnetic field lines. The flow speed, at Earth, is around 400 km/s. The shock, at Earth, is some 17 km thick, and located 90,000 km sunward. Bow shocks exist on planets with a magnetic field, and have been observed in other star systems.

Based on one data point, the only life forms that we are aware of are...us. All the animals and plants. It we see an exo-planet of roughly the size of the Earth, orbiting a star much like our Sun, at a comparable distance, we might conjecture that the other planet might harbor life, like us. If the other planet is not like Earth, we haven't rules out life, just life that we are familiar with.

Near Earth Objects

Technically, an NEO is a solar system object whose closest approach to the Sun is 1.3 AU, and that comes in close proximity to the Earth There are 14,000 known asteroids in this category, 100 comets, solar orbiting spacecraft, and meteoroids. All these have the potential of striking the Earth. They are closely tracked from the ground, by NASA's Planetary Defense Coordination Office. A joint US/EU project called Spaceguard is tracking NEO's larger than 30 meters. Three NEO's have been visited by spacecraft.

Near-Earth Objects (NEO's), there more than 15,000. Stuff hanging around Earth. Chunks of rock. If these enter the atmosphere, they heat up and burn. Sometimes, enough is left to hit the ground, appearing as a rock. The easiest place to find meteors (as we call asteroids that hit the ground) is Antarctica, where they stand out against the snow and ice. A lot of the Antarctic meteors come from Mars, as one of my old professors proved. Specifically, Martian Meteorite ALH84001 found in Antarctica. Can we terraform a NEO? Maybe. They are rather small. Probably can;t hold onto an atmosphere.

The Moon

The next logical step beyond Earth orbit is the Moon, as a destination. There are several options, with varying complexity. Good points: lower gravity, reasonably close, lava tubes, water ice; bad points: temperature extremes, radiation, no atmosphere.

Technically speaking, human have lived on the moon, briefly, using their Apollo lander (LEM) as a base. As I work on corrections and updates for this book, the Artemis missions to go back to the Moon are close to launch.

NASA's Lunar Outpost on the surface project is on hold, but they are pursuing the Deep Space Gateway, essentially and ISS around the Moon. This will allow operation of telerobotic rovers on the surface.

One ideal structure for lunar bases (that the author has worked on) is lava tubes. These are found on the Earth as well, and the cooled lava provides a hard, sealed surface. It just needs to be capped with airlock doors, and no further exterior construction is required. The ceiling thickness is 10's of meters, providing protection from radiation, and impact events. It has lots of advantages, if you don't mind living in a tube. Many lava tubes we know about have been imaged by the Lunar Reconnaissance Orbiter.

We could envision vacation destinations on the lunar surface. Here, we operate in about 1/6 of the gravity we are used to, and the water will stay in the designated pool. Where does the water come from? There is thought to be large amounts of water ice in craters at the lunar poles. This is being sought, because it can be converted to hydrogen and oxygen (rocket fuel) by solar electrolysis.

One incredible unique activity that can be supported at a lunar facility is human winged flight. Since our apparent weight will be 1/6 of Earth normal, we can enter a pressurized volume, don wings, and truly fly. The pressurized volume could be sub-surface, or inflatable, on the surface.

Besides defining variations of Earth-sport, we can implement a new one in Lunar gravity – flying. Since our weight will be 1/6, this is quite feasible with strap-on wings. It will take a bit of getting used to, but look forward to it at the Lunar Icarus Center.

ESA has plans for a lunar resort called the Moon Village. The infrastructure would be launched to the lunar surface and constructed by telerobotics. Besides tourism, the established base could support lunar surface research and manufacturing. The projected date for this project is 2030. It is projected that this facility will be at the lunar poles. Theme Parks, we need lunar theme parks.

The Asteroid Belt, dwarf planets, and Centaurs

Asteroids have been imaged by the New Horizons spacecraft, on its way to Pluto, and by the Cassini spacecraft. The Pioneer-10 spacecraft was sent to study the far reaches of the solar system It passed through the Asteroid belt on its way to Jupiter and Saturn.

Although there are fewer than 10 planets, and less than 200 moons, there are millions of asteroids, mostly in the inner solar system. The main asteroid belt is between Mars and Jupiter. Each may be unique, and some may provide needed raw materials for Earth's use. There are three main classifications: carbon-rich, stony, and metallic.

The physical composition of asteroids is varied and poorly understood. Ceres appears to be composed of a rocky core covered by an icy mantle, whereas Vesta may have a nickel-iron core. Hygiea appears to have a uniformly primitive composition of carbonaceous chondrite. Many of the smaller asteroids are piles of rubble held together loosely by gravity. Some have moons themselves, or are co-orbiting binary asteroids. The bottom line is, asteroids are diverse.

It has been suggested that asteroids might be used as a source of materials that may be rare or exhausted on earth (asteroid mining) or materials for constructing space habitats or as refuelling stations for missions. Materials that are heavy and expensive to launch from earth may someday be mined from asteroids and used directly for space manufacturing. Valuable materials such as platinum may be returned to Earth for a profit.

Except for Ceres, there are not many options for a base of operations, and Ceres is too small to be properly terraformed. If we eventually need a semi-permanent habitat to support, for example, mining missions, we could use Ceres. This could be based on our model of ocean-based oil drilling platforms. The stay time would have to be longer, due to the travel times involved.

There are only 8 ½ planets, but there are thousands of asteroids, and it seems there may be as many types. This means that exploring the known asteroids is a daunting challenge. On the other hand, the asteroids can be a significant source of raw materials for Earth. A conventional survey and exploration approach would take too long. What is needed instead is a multitude of autonomous and flexible nano-spacecraft. The architectural model is a swarm (social insect model) with distributed intelligence.

The asteroids are not uniformly distributed. In the asteroid belt, the Kirkwood gaps are relatively empty spots. This is caused by orbital resonance of the asteroids with Jupiter. Orbiting irregular shaped bodies is challenging, due to the irregular gravity field. This makes station keeping and attitude control a problem.

Centaurs are icy minor planets between Jupiter and Neptune, and there are 3 known. There may be 44,000 others. Sometimes, they are captured by a planet's gravity, as a moon.

The dwarf planets of our solar system include Ceres, Orcus, Pluto, Salacia, Varuna, Haumea, Quaoar, Makemake, 2007 OR10, Eris, and Sedna. These smaller objects did not make the size cut to be a real planet. These all orbit the Sun. Orcus is a trans-Neptunian object, Salacia, Haumea, Quaoar, Makemake, and Varuna are Kuiper Belt objects. Eris is the largest of the dwarf planets, having its own moon. Sedna is beyond the Kuiper belt. It's orbital period (year) is 11,400 Earth years. It's in a highly elongated orbit, probably due to Neptune's gravity. Generally, a dwarf planet does not have enough gravity to clear its orbit of other material. Not all dwarf planets have yet been discovered or observed. There may be 10's of thousands.

Comets

There are some 5,253 known comets. The Deep Impact mission returned images of the surface of comet Borrelly in 2001. That surface was hot (26-70C), dry, and dark. In July of 2005, the same mission sent a probe into Comet Tempel 1. It created a crater, allowing imaging of subsurface material. Water ice was seen. Comet Borrely has a coma, which proved to be vaporized subsurface water ice. Deep Impact went on to complete a flyby of Comet Hartley-2 in 2010. Comets are not going to be a good target for Terraforming. They are predominately frozen gas, that melts as they get closer to the Sun.

Mars

Mars, and its two tiny moons and seven Trojans has got some infrastructure in place – a communications relay and a weather satellite. There are several Rovers and landers on the surface.

The Viking program was a pair of spacecraft sent to Mars in 1975. Each spacecraft consisted of an orbiter, and a lander. A major target now is a Mars sample return mission. The Mars launch window comes along every 780 days.

Mars is, with a stretch, Earth-like. Being smaller, it has less surface gravity. It has no magnetic field, so your compass is useless, and there are no equivalents to the van Allen belts that shield Earth from charged particles from the Sun. Constructing an artificial equivalent of

Earth's van Allen belts has been speculated, or perhaps a big magnetic shield at the Mars-Sun L1 Lagrange point.

Current missions are search for traces of life, or history of life. It seems to have most of the right elements present. There is evidence of ancient surface water, some of which boiled off into space, and some may be found in underground aquifers. Mars' rotation rate and orbital period are similar to Earth, meaning the day and year are similar. It is just within the habitable zone of our solar system.

The habitable zone, sometimes called the Goldilocks" zone is not too hot, not too cold, but just right for life. The planet has to be the right distance from it's "Sun", and have the correct atmospheric pressure to keep the water from evaporating. This definition supports the water-dependent life as we know it.

Mars may have been habitable in the past. It may also harbour life that we haven't found yet. One downside is, the atmosphere is not very dense, so liquid water does not exit on the surface. Sub-superface water ice has been found, and both poles are covered in water ice. Mars has about the same surface area as Earth's continents, since it has no surface water.

Mars has lava tubes around the big volcano, Olympus Mons. There are several possible cave entrances on the volcano Arsia Mons. These might be a good location for a first colony. They would provide protection from radiation, excessive UV light, and micro-meteors. Sealing the entrances would allow for a near Earth-normal environment.

Let's fast forward a few years where terraforming hasn't gotten much of a start, but colonization is an option. We would need a colony that manufactures most of its needs onsite, and doesn't require frequent logistics flights. This means a local source of water and oxygen, and greenhouses for local food production. With local water ice, we can get oxygen (and hydrogen, for rocket fuel).

The surface soil has been analysed in-situ, and seems to have the right elements to support plant life. The surface gravity is about 38% of Earth's. It has a dense core. The dust on the surface is mostly iron oxide, but traces of other elements as well. There are impact craters, as its thin atmosphere is useless in burning up meteors.

Most of Mars' scanty atmosphere is carbon dioxide. There is a trace of methane, which is useful for rocket fuel, and as a carbon source. Methane can come from natural sources such as volcanos, or from microbial action. A prototype facility is being tested, at KSC that could manufacture water, oxygen, and rocket fuel from Martian soil. The soil can give us water, that is broken down into hydrogen and oxygen. Hydrogen is hard to store, but we can combine it with carbon to make methane, which is CH_4. The carbon comes from Mars' atmosphere's carbon dioxide. Methane makes a good rocket fuel. We might also just use the hydrogen and oxygen, if we can keep them cold enough.

At this writing, Mars has fifteen spacecraft in orbit, and at least 3 functional on the surface, with an additional 4 inactive. Several other landers and rovers are now awaiting the construction of the Mars Museum.

NASA, ROSCOSMOS, ESA, ISRO all have existing and planned Mars missions. Private space Company Space-X is planning a tourist trip to Mars. The U.S. Senate directed NASA in 2017 to "get humans near or on the surface of Mars by the early 2030's."

A specific program addressing the Mars issues is the Flashline Mars Arctic Research Station (FMARS). There is currently one such facility in the Arctic, with a second in the southern American desert. The first station is on Devon Island in the Arctic sea. It is located on a ridge, overlooking a large impact crater, about a thousand miles from the North Pole. The facility was built in 2000, and is operated by the Mars Society, a non-profit. It is used to define and refine field procedures, test habitat design, study crew performance, and selection criteria. It began operations in 2001. Generally, there is a core crew of ten, with visiting researchers and assistants. Communication to and from the station to external sources is delayed 20 minutes, to simulate the one-way radio/light travel time to Mars. The crew keeps to the somewhat longer Martian Sol day.

The Biosphere-2 project, located in the Arizona desert, supported 8 humans for a year in a closed ecosystem. It is rather large, covering more than 3 acres. It operated twice, in 1991-93, and in 1994. It was built the Space Biosphere Ventures in 1987-91. During its operational periods, much good data was compiled. It is now operated by the University of Arizona.

Besides being our first planetary home-away-from-home, Mars could serve as a base for observatories for the outer planets, and stars. It is probably the most "Earth-like" in the sense of what has to be done to make it habitable. The gravity is 38% of Earth normal, but we can't do anything about that. A lower gravity, yet one strong enough to keep things from falling up, is a positive. The atmosphere is about 1% of sea level on Earth, so we need more of that. We might also augment Mars' methane and other hydrocarbons, which are abundant in some of Jupiter's and Saturn's Moons.

There is a lot of CO_2 tied up in the surface regolith and the solar ice cap, and that would be available if we warmed the planet up. Not so much that we got thermal run-away like at Venus. In the mean time, residents could live in habitats, and use simple suits outside. We need nitrogen for the atmosphere as well, and we might get this from asteroid mining – think of sending a big chunk of frozen nitrogen from the asteroid to impact Mars. Carefully.

One suggestion is to darken the soil of Mars to reduce the albedo and allow the planet to warm. Material from the Martian moons could be used, or material from the asteroid belt.

Due to the thin atmosphere, Mars has significant wind storms that create vast, planet-wide dust storms. These affect surface visibility and obstruct. solar panels.

Two big questions are: did Mars ever host life, and, can we live there? Mars doesn't have much of an atmosphere, which means there is a lot of solar ultraviolet light. Good for evolution, bad for life as we know it.

Gas Giants

The Gas giants are the planets Jupiter, and Saturn. Exploring these is the responsibility of the Jet Propulsion Laboratory. These planets each have extensive ring and moon systems that have been imaged, but are just beginning to be explored.

The Gas Giants are large planets, beyond the orbit of Mars. They consist mostly of the gases hydrogen and helium. At depth, they consist of liquid metallic hydrogen. There may be a core of solid hydrogen, or a rocky material. It is possible these elements exist as a liquid at extreme depths. Essentially, a gas giant planet is a failed star. The nuclear fusion process did not get started.

The gas giants have prominent cloud systems in the upper atmosphere. Extrasolar Gas giants have been discovered orbiting other stars than our Sun.

Jupiter

Jupiter has 79 known moons, and perhaps 1 million Trojans of 1 kilometer or larger. These tend to congregate at the L4 and L5 Lagrange points. The largest has a diameter of several hundred kilometers. The International Astronomical Union just announced as this book was being updated the discovery of 12 previously unknown moons of Jupiter, by an observatory high in the Andes in Chile. Only one has been named so far, Valetudo, a great-granddaughter of Jupiter. Jupiter has a ring system with three main segments.

Jupiter has a very high trapped radiation environment. The moons are mostly all different, and some are thought to be capable of hosting life, as we know it. The moon Io has volcanic activity, as seen by the current Juno mission. The Gas Giants do not fit our profile for terra-form-able planets. However, the interesting and extensive moon systems do. There's quite a bit of diversity, some being "water-worlds", some with a sub-ice ocean, some rocky.

Jupiter has a massive, persistent cloud cover, mostly of ammonia, some 50 km thick. There are winds gusting to 350 km/hour and turbulence. Jupiter has a magnetic field, some 14 times stronger than Earth's. Jupiter itself may not prove to be a good spot to put down roots, bu tits moon certainly are.

The moon Europa has water ice on the surface. It is considered "one of the most promising extraterrestrial habitable environments in our solar system" according to the most recent Planetary Society's Decadal Survey. A proposed mission, ExCSITE, would provide characterization of the surface properties.

The four largest of Jupiter's moons can be seen from Earth with a modest telescope. They could qualify as dwarf planets, if they orbited the Sun, not Jupiter. The rest of the moons, and we have probably not found them all, are irregular, orbiting randomly. For terraforming candidates, we have the Galilean moons, Io, Europa, Ganymede, and Callisto.

Io is the 4th largest of the Galileo group, and has the highest density. It has the least water of any body in the solar system, seen so far. Over 400 volcanos have been observed. The crust of the moon seems to be silicate and iron. Due to gravitational forces from Jupiter and the other 3 Galilean moons, Io gets jerked around a lot, leading to tidal heating, and a lot of volcanism. Not our top candidate.

Europa is the smallest of the 4 Galilean moons, slightly smaller than Earth's moon. It has a water-ice crust, and most likely, an iron-rich core. It has a very smooth surface. It is possible that the moon has a sub-surface ocean, which is warmed by tidal flexing. Europa is a good candidate for consideration of Terraforming. It would serve as a good outpost for further investigation of the Jovian system, and perhaps as a base for ring mining.

Almost all of the moons are in tidal lock with Jupiter, meaning the same side always faces the planet.

Ganymede has the distinction of being the largest moon in our solar system. It is a bit larger than the planet Mercury, but doesn't seem to have much of an atmosphere. It has a metallic core, and a magnetic field. Its composition is mostly silicate rock and water ice. It is estimated to have more water than Earth. There is a thin oxygen atmosphere. Pretty close to what we're looking for,

Callisto is also a possibility, and it is further from Jupiter and its trapped radiation belts. Colonies at these distances from Earth would certainly need to be self-sufficient, due to the long travel times involved.

We're going to have to know a lot more about the Jovian system before we decide to establish an outpost there.

Saturn

Saturn and it's 62 known moons has a one-way light time around 1.4 hours. Saturn has been visited by spacecraft four times. The first was a flyby by Pioneer-10 in 1979. This showed the temperature of the planet was 250 degrees K. Voyager-1 visited in 1980. It conducted a close flyby of the moon Titan to study its atmosphere. It is, unfortunately, opaque in visible light. Voyager-2 swung by a year later, and data showed changes in the rings since its sister mission visited the year before.

Temperature and pressure profiles of the atmosphere were gathered. Saturn's temperature was measured at 70 degrees above absolute zero at the top of the clouds, and -130 C near the surface. The flybys discovered additional moons, and small gaps in the rings. Like Jupiter, we will probably establish a facility on one or more moons.

ESA's Cassini was the fourth spacecraft to study Saturn, which has rings, although smaller than Jupiter's. The rings were confirmed by the Voyager spacecraft in the 1980's. Cassini entered into Saturnian orbit, and returned much valuable data. At the end of mission, it was de-orbited into Saturn, and burned on entry into the atmosphere. This was to preclude any biological contamination from Earth. The film of it entering the atmosphere was nominated for an Emmy award. It has also collected data on the Saturnian moons Titan, Enceladus, Mimas, Tethys, Dione, Rhea, Iapetus, and Helene. The one-way communications time varies from 68-84 minutes.

Things are strange in the Saturnian system. Cassini observed a hurricane in 2006 on the planet's south pole. It appears to be stationary, 5,000 miles (8,300 km) across, 40 miles (67 km) high, with winds of 350 mph (560 kph). The large moon Titan has lakes of a liquid hydrocarbon, with possible seas of methane and ethane. The Cassini mission was responsible for the discovery of seven new moons of Saturn. It conducted a close flyby of the moon Phoebe, and two fly-bys of Titan. Phoebe has a retrograde orbit. Cassini launched the Huygens probe onto the moon Titan. It landed, and continued to supply data. Cassini was also responsible for discovery of another ring, as well as finding lakes of hydrocarbons near the north pole of Titan. These turned out to be larger than first thought, and were renamed seas. Keeping busy, Cassini discovered four additional moons in 2009. Lasting beyond its primary mission of completing 74 orbits, the spacecraft went into extended missions, operating until 2017, when it entered the Saturnian atmosphere.

The rings of Saturn are the most extensive, larger than Jupiter's. They consist mostly of water ice. The astronomer Huygens was the first to observe and describe them in 1655, subsequently getting a spacecraft mission named after him. There are two gaps in the rings, caused by moons orbiting at that distance from the planet. Orbital resonance by a dozen of Saturn's moons cause other gaps. The outermost moon, Phoebe, is tilted at 27 degrees to the others, and orbits the planet in retrograde, opposite of the other rings and satellites. The Cassini division (named after its discoverer) separates the inner B ring, and the outer A ring.

The moon Enceladis was observed to exchange plasma waves with the planet, and the ring system. Cassini's Radio Plasma Wave Science (RPWS) instrument recorded intense plasma waves during one of its closes encounters. The moon is within the Saturnian magnetic field. It is also geologically active, sending out plumes of ionized water.

The moons Titan and Enceladus are outside the habitable zone, but do have liquid water in subsurface oceans. Titan may have lakes of hydrocarbons, based on the Cassini fly-by data. The Cassini mission discovered a possible atmosphere on the moon Enceladus, along with with ionized water vapor, and ice geysers. Many of the Saturnian moons are in tidal lock with their mother planet. Being so close to its giant neighbor Jupiter affects the Saturnian system.

Cassini observed a massive storm on Saturn, the great white spot, that recurs every 30 years. The storm, larger than the red one on Jupiter, exhibited a discharge that spiked the temperature 150 degrees. At the same time, Earth observations showed a large increase in atmospheric ethylene gas. It also discovered large lakes or seas of hydrocarbons near the planet's north pole

Ice Giants

Uranus and Neptune differ from Jupiter and Saturn in that they have heavier volatile substances in their atmosphere, and are now referred to as ice giants. Their atmospheres are known to contain water, ammonia, and methane ice. Again, the planets themselves would take a lot of work and time to make them more like home, but some of their moons may be suitable.

Uranus

Interestingly, Uranus' rotation axis is tilted into the plane of its orbit around the Sun. Seasonal changes and weather have been observed. The Voyager-2 mission imaged Uranus on its way from Jupiter, and out of the solar system. Atmospheric wind speeds are know to approach 900 kilometers per hour. It's orbital period is 84 Earth years. It receives about $1/400^{th}$ of the light that the Earth does from the Sun, so solar power is not a viable choice. In addition, the moons rotate in the plane normal to the velocity vector. Uranus is sideways, or lying on its side, compared to all the other planets. It's spin axis is 98 degrees out of the orbital plane with the Sun.

The ring system was not discovered until 1977, since they are dark and faint. Nine were know until the Voyager visit, when an additional two were discovered. The Hubble Space Telescope found an additional 2.

Because of the strange orientation of the planet's rotation axis, during the solstice, one side of the planet faces the Sun continuously, and the other faces deep space. Each pole gets 42 years of direct (though weak) sunlight, and 42 years of darkness. In spite of this, the equator is the hottest region. At this writing, the planet is in its autumnal equinox. Uranus does has a magnetic field.

Uranus has thirteen inner moons, five major ones, and nine irregular moons. The inner moons have the same properties are the ring system. The major moons show signs of volcanism. The irregular ones have strongly inclined orbits, some retrograde. The major moons are roughly in the plane of the planet's equator.

Five moons would be considered planets if they orbited the Sun, not Uranus.

The moon Titania is Uranus' largest, and probably has a rocky core, with surface ice. It is about one half the size of Earth's moon. There may be liquid water under the ice. The surface shows impact craters, There is frozen carbon dioxide on the surface. It is possibly a good target for terraforming, and perhaps a base to observe Uranus more closely.

The moon Oberon is the outermost major moon also probably has a rocky core and an icy covering, with liquid water at the boundary. This would require ammonia or an other anti-freeze to be present. There is evidence of asteroid impacts on the surface. We're not writing it off, right now.

Ariel

This moon seems to be made up of water ice and a dense, non-ice element such as rock. The surface ice is crystalline, and there is evidence of carbon dioxide. Some of the surface is cratered.

Umbriel

The composition of Umbriel is similar to that of Ariel. It also has visible impact craters.

Miranda

Miranda is the smallest of the five round satellites, with a diameter of 475 km. It has a very low density, that may be due to a 60% composition of liquid water. It has a rocky surface. Water has been detected on the surface.

Uranus was visited by Voyager-2 in 1986, after the spacecraft went by Jupiter and Saturn. It discovered ten moons, and two additional rings in the known ring system. Voyager also captured some images of the Uranian moon Umbriel, and discovered a magnetic field. Since the planet is tipped 90 degrees, the magneto-tail is cork-screw shaped. The ring system is fundamentally different from that of Jupiter or Saturn.

Uranus itself has a strange predominately water-ammonia ocean, which is electrically conductive. A future planned mission is the Uranus Orbiter and Probe. There is a planned launch in 2030, with Uranus orbit insertion in 2041. The mission is named Oceanus. (Origins and Composition of the Exoplanet Analog Uranus System).

Neptune

Neptune has 14 known moons, and 18 known Trojans.

Neptune has also been visited by Voyager-2 in 1989. It discovered six new moons. That is the extent of close-up observations of the planet. Neptune has rings, like Jupiter and Saturn, and a great dark spot. It's moon Triton has geysers and polar caps.

Neptune was visited by the Voyager-2 mission in 1989, when six moons were discovered. That is the extent of close-up observations of the planet. Neptune has rings, like Jupiter and Saturn, and a Great Dark Spot in the Southern hemisphere. It's largest moon, Triton, has geysers and polar caps. The moon Triton has an interesting retrograde orbit – it goes in a different direction than the other moons. Triton's surface is mostly frozen nitrogen, and is geologically active. It is speculated that Triton has a subterranean ocean. The moon Ptoteus is an ellipsoid, not a sphere. The atmosphere is mostly hydrogen and helium, with some hydrocarbons. It also has water, ammonia, and methane ice. Neptune also has a significant magnetosphere, that interacts with the solar wind. Not too many good candidates here.

Now that Pluto has been downgraded to minor planet status, Neptune is officially the farthest planet in our solar system from the Sun. It is at an average distance from the Sun of 30 AU, and orbits every 165 years. It was found by mathematical prediction, as it is not visible without a good telescope. Something was perturbing the orbit of Uranus. The intruder was predicted in a certain location, and was then observed by telescope in 1846. In

Galileo's notes from 1612-13, there are points that correspond to the position of Neptune. His telescope was just not quite good enough to resolve it. He mentioned that the object moved relative to the "fixed" stars.

By spectroscopic analysis, the atmosphere is mostly of hydrogen and helium, with some hydrocarbons, and trace amounts of nitrogen. At the equator, methane, ethane, and acetylene are found. It has a blue color, due to methane in the atmosphere. It's atmosphere has visible weather patterns, with winds of up to 1,300 mph.

Neptune has a magnetosphere, with a magnetic field tilted some 47 degrees from the spin axis. It has extreme weather, with winds reaching supersonic speeds. No good candidates here, except possibly for Triton. More data is needed.

Pluto and beyond

Pluto was downgraded from a planet to a Kuiper Belt object. The *New Horizons* mission to Pluto and the Kuiper Belt began in January of 2006, and reached the vicinity of Pluto in July 2015. It conducted a 6-month survey of Pluto, and went out farther into the Kuiper belt, on an 3 year extended mission, which is ongoing at this writing. The spacecraft was developed for NASA by the Johns Hopkins University Applied Physics Lab in Laurel, MD.

Pluto had one known moon, Charon, before the New Horizons Mission Team members, using Hubble Space Telescope data, discovered four more, Nix, Hydra, Styx, and Kerebos.

There is lots of interesting stuff beyond Pluto, before Inter-stellar space is reached. The minor planet 90377 Sedna is located at about 86 AU and is one of the Trans-Neptunium Objects. These are too far away at the moment to consider.

The Scattered Disk is a group of objects in elongated orbits, caused by the influence of Neptune. It extends from 30 AU to about 100 AU. Located in the Scattered disk, the dwarf planet Eris has the distinction of being the most massive known dwarf planet. It does orbit the sun, and is classified as a Trans-Neptunian Object. It is slightly more massive than Pluto, yet smaller. Data are scanty, but it may have a liquid water ocean. Methane ice has been spotted. A flyby mission from Earth would take some 25 years to get to target.

The Kuiper belt is a circumstellar disk, extending from Neptune (30 AU) to 500 AU. It resembles the asteroid belt. It holds mostly small bodies.

The Oort cloud is a circumstellar disc around our Sun, containing icy worlds. It extends from 2,000 to 200,000 AU. There is a disk-shaped inner cloud, and a spherical outer cloud. These, strictly speaking, do not belong to our solar system, but are located in interstellar space. It is the domain of comets.

Challenges

Once we leave the vicinity of our home planet, conditions deteriorate quickly. The major issue is radiation, since we are outside of safety of the trapped radiation belts, which provide some protection. This is the major challenge, but there are many known ways to mitigate this problem. Then, there is the thermal problem. We're going somewhere that's hotter (sun-ward), or colder. A big issue is the mission duration. It takes years to get to some of the outer planets. Missions outside the rather friendly environment of near-Earth face additional challenges that must be addressed.

These are all solvable problems, but require additional engineering analysis, and the development of new technologies.

Power

Out to the orbit of Jupiter, solar cells work well for exploration missions. Beyond that, a large spacecraft will not get enough electrical power. A traditional approach has been to use RTG's – Radioisotope Thermoelectric Generators. These work by using thermocouples to convert the heat from the decay of a radioactive material into electricity. No moving parts. There is a safety concern during launch that the radioactive material could get scattered in an explosion. The units are good for a few hundred watts of power, for decades.

A common radioactive source is plutonium-238, with a

half-life of 88 years. RTG's have been used in space since 1961.

Due to advances in solar cell efficiency, the Juno mission to Jupiter successfully used them for power, in lieu of a RTG. The Voyager spacecraft, in interstellar space, continue to transmit back data, and will have enough power from their RTG's for a century or so. Smaller radioactive heater units are also used for long distance missions.

Radiation Environment

The radiation environment once you get past Mars is less influenced by the Sun, then by Cosmic rays. Of course, Jupiter, with its strong magnetic field, has managed to accumulate significant trapped radiation belts, from its interaction with the solar wind. In transit, the main radiation issue is energetic cosmic rays.

Energetic protons are also a problem, as are gamma rays. These can cause single-event upsets in electronics, as well as cell damage to humans. Cosmic radiation originates outside the solar system. It consists of high energy protons, and atomic nuclei. It is postulated that Cosmic rays originate from stars' supernova explosions.

Exo-challenges

As difficult as forcing another planet in our solar system to conform to our norms, consider how difficult that would be for a planet of another star, an Exo-planet.

Quite a few of the roughly 4,000 exoplanets we know of are "Earth-like", meaning the same size, and roughly the same distance from their star, as Earth. Of course, there are the distance challenges. By the time we are ready to travel interstellar, we should have a good idea of where we can go, that is a lot like home. At that point, we may find our target already inhabited.

We have to be very careful in attempting to make another planet like Earth. We don't want to make a non-reversible mistake. Colonists could live in a habitat without Terra-forming, while the planet is studied. Any change we kick off will take a long time to happen, decades, if not centuries.

On Earth, we have tried analogs to habitats, particularly for Mars. These have taught us a lot, but things have not always worked out well. The Biosphere-2 project failed when an un-anticipated chemical reaction in the concrete removed the oxygen from the air. That's not a survivable event on another planet.

We would like to have roughly the same length of day as we have on Earth, and a similar or lesser gravity. We need a certain temperature range, and humidity. We also need a Plan-B, a lifeboat, when all else fails.

If we kick off a Terra forming project, we are going down the road of changing a planets biosphere. Are we sure there was no life there, that we're going to wipe out? There are planetary protection protocols in place now that, for example, dictate sterilization of surface landers. This does NOT eliminate all Earth organisms. The spacecraft can have no more than 300,000 spores on the exterior. We have already contaminated Mars. We might actually have a mini-terraforming project going on at Mars now. Terra-forming will be a slow process, but we probably can't stop it, once we kick it off. And, Earth life is well evolved to living on Earth. It could be the basis for life whatever planet we choose to tackle, but it will necessarily evolve to survive and flourish its new environment.

There will be an incredible cost to Terra-forming, one we can probably never fully recover. But we weigh this against extinction if something natural or that we screw up effects the Earth. Exploitation of local resources might offset some of the cost. We wouldn't need two diamond mountains, after all.

If we're in a habitat, observing and "controlling" the process, there will necessarily be contamination, because the habitat will have small leaks, as will the space suits. We also have to realize that we will not get a clone of Earth. Too many environmental parameters are different. Only the Earth is "Earth." On Mars, we are all Martians. One handy thing is, we can "mine" the asteroid belt for materials to use on Mars.

A good Mars analog is Antarctica. It is agreed upon among nations it is a science preserve, and there is no commercial activity. The average temperatures are about the same. There are people living there all year, although most science takes place during the summer. Almost everything has to be carried in (and waste material carried out). It takes a while to get there, although the journey is nothing like that to Mars. There is, by the way, a greenhouse at McMurdo Station, providing fresh veggies.

As our Terra-forming Project goes along, we will need to establish biodiversity, having multiple species that co-exist, and benefit each other. We also have to remember that evolution is a continuous process. Due to different environmental drivers, Earth species transferred to another planet will evolve differently than those back on Earth.

Wrap up

What's next? Well, when it starts getting too crowded at home, or we need more resources, we could head out to our nearest neighbor, the binary star system, Alpha Centauri, some 4.37 light years distance. If we launched a probe today, it could reach our closest neighbor Alpha Centauri in 40,000 years. We would really like to get some data in our lifetimes. NASA has discussed an interstellar mission for 2069.

The next step beyond a remote, fixed colony on another planet is a colony ship, which would be on a long journey. Various approaches have been looked at, such as the generation ship, which is the home to multiple generations of space-farers, and the sleeper ship, which uses hibernation.

From history, the *Mayflower* was a colony ship, carrying some 100 people and what they would need for a while in a lush, yet underdeveloped land in the 1600's. There were 102 passengers, in an area of some 25 by 15 feet. Cargo included goats, pigs, and poultry. The first winter after arriving, the ship served as a habitation for the colonists. Roughly half of them died.

The Mayflower returned to England, which is something an intersteller colony ship probably wouldn't do. It would instead stay on site, as temporary housing for the colonists. No colonists returned to England, only the ships crew.

The analogy is stretched. The Mayflower colonists expected and found an environment that they could live in. Interstellar colonists may not. Mayflower-II is not a bad name for a joint UK-US interstellar mission. Can we expect friendly residents awaiting us, willing to share their land? Let's not bet on that.

Afterword

Terra-forming will take place, but the stakes are high. First, we may need to make some tweaks in Earth's environment, and we have to get that right the first time. Then, we should choose a near-by planet, Mars, Venus, to try out our techniques to manufacture a "New Earth." At some point, permanent habitability on these second homes will be feasible without costly logistics flights. Hopefully, we will be dispersed widely enough to avoid any catastrophe. It's a big solar system out there. Let's tidy it up a bit, and head out to another one as well.

Glossary of Terms

Aerobe – organism that can grow in an oxygenated environment.
Aerostat – tethered lighter than air platform.
Albedo – *Whiteness*, in Spanish; amount of sunlight reflected.
Anticyclonic – counter-clockwise rotation.
Apojove - the point furthest away from Jupiter by a body that orbits it.
Apsis – extreme point in an orbit.
ASIN – Amazon Standard Inventory Number.
Asteroid - minor planets, orbiting the Sun.
AU – astronomical unit, mean distance from the Earth to the Sun
BEO – beyond Earth orbit.
Biodome – a closed ecosystem.
Bow shock- Where the solar wind begins to interact with a planet's magnetosphere.
BTO – Biological Technological Office (DARPA).
CCSDS – Consultative Committee for Space Data Systems, standards organization.
Centaur – a minor planet in an unstable orbit, behaving like an asteroid or comet.
CHZ – circumstellar habitable zone
Comet – icy body orbiting the Sun inn a very eccentric orbit.
CDF - (ESA) Concurrent Design Facility.
ConOps – concept of operations.
DARPA – (U.S.) Defense Advanced Research Project

Agency.
Death zone – atmosphere below 5.4 psi.
DLR – German Aerospace Center.
DRM – design reference mission.
DSN – Deep Space Net
Ecliptic plane – the apparent path that the Sun seems to follow, the same as the Earth's orbit.
ECLSS – Environmental Control & Life Support System.
ESA – European Space Agency.
EXOMars – Exobiology on Mars (ESA/Russia)
Galilean Moons – the four moons spotted by Galileo – Io, Europa, Ganymede, Callisto
Gas giant – A large planet consisting mostly of hydrogen and helium.
Goldilocks zone – habitable zone around a star, distance where liquid water can exist on the surface. No too hot, not oo cold
Great Red Spot – a very large storm on Jupiter, larger than Earth.
GSFC – NASA Goddard Space Flight Center, Greenbelt, MD.
HAVOC – High Altitude Venus Operational Concept (NASA).
HED - class of meteorites, howardite–eucrite–diogenite.
Hypoxia - inadequate oxygen.
HOPE – human outer planet exploration.
Hydrosphere – the amount of water a planet has.
IAU – International Astronomical Union.
Ice giant – A large icy/liquid planet, consisting of elements heavier than hydrogen and helium.

ISBN – international standard book number.
ISRO – Indian Space Research Organization.
ISRU – in situ resource utilization.
ISS – International Space Station.
JOI – Jovian orbit insertion,
Jovian – pertaining to Jupiter.
JPL – Jet Propulsion Lab, operated by Cal Tech for NASA;   responsible for all deep space missions.
JSCV – NASA's Johnson Space Center, Texas.
Katabatic winds – winds from a higher elevation.
KBO – Kuiper Belt Object.
Kph – kilometers per hour.
KSC – Kennedy Space Center.
Kuiper Belt – beyond Neptune, a ring of small icy asteroids and minor planets.
Lagrange (L) point – a null point in the gravity field in the 3-body program.
    L1 - the Lagrange point between the 2 bodies.
    L2 – the Lagrange point behind the smaller body.
    L3 – the Lagrange point behind the larger body.
    L4 – the leading Lagrange point in an orbit.
    L5 – the trailing Lagrange point in an orbit.
LGM – little green men.
LEM – Lunar Excursion Module (Apollo).
LSSPO – (NASA) Lunar Surface Systems Project Office, JSC.
Magnetosphere – a space surrounding a planet or moon that is affected by the primary's magnetic field.
Magnetopause – abrupt boundary between a magnetosphere and the solar wind.

Magnetotail – magnetosphere extends away from the planet and the Sun.
MARPOST – Mars Piloted Orbital Station (Russia).
Moon – object in orbit around a planet.
Mph – miles per hour.
MSFC – Marshall Space Flight Center.
MSL – Mars Simulation Lab, (DLR).
NASA - National Aeronautics and Space Administration.
NEO – near Earth object.
NIAC – NASA Institute for Advanced Concepts.
One-way light time – a measure of distance, in terms of how long it would take light to travel the distance.
Orbit – the path of one body around another, that are linked by gravity.
Planet - object in orbit around a star.
Rassor – Regolith Advanced Surface Systems Operations Robot.
Regolith – loose soil, covering rock. Dirt.
Ring system – a disk of solid material around a planet.
Roche's limit – withing 2.44 radii of the planet, no stable moon is possible, due to tidal forces from the primary.
ROSCOSMOS – Russia Space Agency
RTG – Radioisotope Thermal Generator – electrical power plant.
Selenology – geology of the moon.
SHFE – space human factors engineering
SI – System International – the metric system.
SOI – Saturn Orbit insertion; sphere of influence.
Solar flare – a sudden rapid emission of electrons, ions,

and atoms from a star.
Solar System – A star and its associated planets and such.
Solar wind – stream of charged particles emitted from a star's upper atmosphere.
SNAP – Systems for Nuclear Auxiliary Power (RTG).
Statite – static satellite; using a solar sail to hover.
Terra-forming – modifying an environment to be Earth-like, making it suitable for human habitation.
TNO – Trans-Neptunian objects.
Trojan - minor planet that shares an orbit with one of the larger planets.
TRL – Technology Readiness Level.
USGS – United States Geological Survey.
V&V – verification and validation.
VIM – Voyager Interstellar Mission.
Vivarium – artificial ecosystem for research.

Bibliography

Beech, Martin *Terraforming: The Creating of Habitable Worlds*, 2009, Springer, ASIN-B00BWZ7AW8.

Berton, Pierre *The Klondike Fever: The Life And Death Of The Last Great Gold Rush,* 2015, ASIN-B06XGD1TCX.

Brown, Laurie *Recent Terrains: Terraforming the American West*, 2000, JHU Press, ISBN-0801864003.

Cichana, Timothy; O'Delb, Sean; Richey, Danielle; Bailey, Stephen A.; Burche, Adam "MARS BASE CAMP UPDATES AND NEW CONCEPTS," 2017, IAC-17, 68th International Astronautical Congress (IAC).

Coombs, Cassandra R.; Hawke, B. Ray "A search for intact lava tubes on the Moon: Possible lunar base habitats," The Second Conference on Lunar Bases and Space Activities of the 21st Century, NASA. Johnson Space Center, Sept. 1992. Bibcode:1992lbsa.conf..219C.

Eckhart, Peter *The Lunar Base Handbook,* 1999, 1st ed, McGraw-Hill Primis Custom Publishing, ASIN-B01A1MSBRK.

Fogg, Martyn J. *Terraforming: Engineering Planetary Environments*, 1995, ISBN-1560916095.

Häuplik-Meusburger, Sandra Olga Bannova, Olga Space Architecture Education for Engineers and Architects: Designing and Planning Beyond Earth (Space and Society), 2016, Springer, ISBN-9783319192796.

Heppenheimer, T. A. *Colonies in Space, A Comprehensive and Factual Account of the Prospects for Human Colonization of Space*, 1977, ISBN-0-446-81-581-0.

Kaku, Michio *The Future of Humanity: Terraforming Mars, Interstellar Travel, Immortality, and Our Destiny Beyond Earth,*, 2018, ASIN-B07173C875.

Kirk, Charles *The Terraforming of Mars*, 2017, ASIN-B078LTGJRW.

Häuplik-Meusburger, Sandra Olga Bannova, Olga *Space Architecture Education for Engineers and Architects: Designing and Planning Beyond Earth* (Space and Society), 2016, Springer, ISBN-9783319192796.

Heppenheimer, T. A. *Colonies in Space, A Comprehensive and Factual Account of the Prospects for Human Colonization of Space*, 1977, ISBN-0-446-81-581-0.

Irwin, Patrick G. J. *Giant Planets of Our Solar System: Atmospheres, Composition, and Structure.* 2003 Springer, ISBN-3-540-00681-8.

Joynson, Charles *The Terraforming and Colonization of Venus: Adding life to the Planet Venus,* 2018, ASIN-B078XDN58N.

Joynson, Charles *The Terraforming and Colonization of Mars,* 2017, ISBN-0995674124.

Kaku, Michio *The Future of Humanity: Terraforming Mars, Interstellar Travel, Immortality, and Our Destiny Beyond Earth,* 2018, ASIN-B07173C875.

Kanas, Nick *Humans in Space: The Psychological Hurdles,* Springer Praxis Books,2015 Ed, ISBN-3319188682.

Kirk, Charles *The Terraforming of Mars*, 2017, ASIN-B078LTGJRW.

Kleiman, Matthew J. *The Little Book of Space Law,* 2014, ABA, ISBN-1614388741.

Ley, Willy *Gas Giants: The Largest Planets,* 1970, McGraw-Hill, ISBN-0070376387.

Lighthouse, Richard *Terraforming the Atmosphere of Venus*, 2017, ASIN-B073Z3BM7Z.

Mackenzie, Bruce "To Mars - a Permanent Settlement on the First Mission," presented at the 1998 International Space Development Conference, May 21–25, Milwaukee

WI.

Mayer, Alastair, Mayer *Alpha Centauri: First Landing*, 2016, ISBN-1539132295.

Mackenzie, Bruce "To Mars - a Permanent Settlement on the First Mission," presented at the 1998 International Space Development Conference, May 21–25, Milwaukee WI

Melville, Graeme P. "Lava tubes and channels of the Earth, Venus, Moon and Mars," Department of Physics, University of Wollongong, 1994, avail: https://ro.uow.edu.au/theses/2859.

Mendell, Wendell W. *Lunar bases and Space Activities of the 21st century,* 1985, Lunar and Planetary Institute, ISBN 0-942862-02-3.

NASA, *NASA Space Technology Report: Lunar and Planetary Bases, Habitats, and Colonies, Special Bibliography Including Mars Settlements, Materials, Life Support, Logistics, Robotic Systems,* ASIN-B00CLX44E2.

NASA, *Human Missions to Mars: Comprehensive Collection of NASA Plans, Proposals, Current Thinking and Ongoing Research on Manned Mars Exploration, Robotic Precursors, ... Science Goals, Design Reference Mission,* 2012, ASIN-B009MTB0OA.

NASA, *Wernher von Braun's 1969 Manned Mars Mission Plans after Apollo and the Boeing 1968 Integrated Manned Interplanetary Nuclear Spacecraft Concept Definition Study, 2012,* ASIN-B009MQUFEY.

O'Neill, G. K. *The High Frontier: Human Colonies in Space*, 2000, ISBN-189652267X.

Oberg, James E. *New Earths*, Stackpole Books 1981; New American Library, 1983. ISBN-0811710076.

Park, Sang-Young, et al "Mission design for Human Outer Planet Exploration (HOPE) using a magnetoplasma spacecraf*t,"* Planetary and Space Science, Volume 54, Issue 8, August 2006, Pages 737-749.

Portree, David S. F. *Humans to Mars: Fifty Years of Mission Planning, 1950–2000,* NASA Monographs in Aerospace History Series, Number 21, February 2001, NASA SP-2001-4521. Avail: ASIN-B014RGH7GM.

Rapp, Donald *Human Missions to Mars: Enabling Technologies for Exploring the Red Planet,* 2015, ISBN-3319222481.

Sagan, Carl; Druyan, Ann *Pale Blue Dot: A Vision of the Human Future in Space*, 2011, ASIN: B004W0I3LW.

Schrunk, David; Sharpe, Burton *The Moon: Resources,*

Future Development and Settlement (Springer Praxis Books), 2007, ISBN-0387360557.

Seedhouse, Erik *Lunar Outpost: The Challenges of Establishing a Human Settlement on the Moon,* 2008, Springer, ISBN-0387097465.

Seedhouse, Erik *Interplanetary Outpost: The Human and Technological Challenges of Exploring the Outer Planets*, 2012, ISBN-9781441997470.

Soloman, Lewis D. *The Privatization of Space Exploration: Business, Technology, Law and Policy, 2011,* ISBN-978-141284756.

Robert Southey, "The Story of the Three Bears", 1837, in *The Doctor.*

Tsiolkovsky, Konstantin E. *Selected Works of Konstantin E. Tsiolkovsky*, 2004, University Press of the Pacific, ISBN-141021825.

Valier, Max; Miller, Ron (ed), *A Daring Trip to Mars*,1928, reprint, 2013, ASIN-B00CSWANK0.

Young, Anthony *The Twenty-First Century Commercial Space Imperative* (SpringerBriefs in Space Development), Springer, 2015, ISBN-331918928X.

Walker, Robert *Trouble with Terraforming Mars: Great if*

it works - but what happens if it goes wrong?, 2015, ASIN-B014ZKJG14.

Walker, Robert *Terraforming Mars - far into Realms of Magical Thinking: Imagine a future Earth with vast capabilities - would they be grateful for our early irreversible attempts?,* 2015, ASIN-B0150YIU70.

Wiktorowicz, Sloane J.; Ingersoll, Andrew P. "Liquid water oceans in ice giants", Icarus 2007, V 186 (2): pp 436–447.

Zubrin, Robert, *Mars on Earth, The Adventures of Space Pioneers in the High Arctic* 2003, ISBN-1-58542-255-X.

Zubrin, Robert, *How to Live on Mars: A Trusty Guidebook to Surviving and Thriving on the Red Planet*, 2008, ASIN-B001FA0K1I.

Jupiter

Bagenal, Fran; Dowling, Timothy E. *Jupiter: The Planet, Satellites and Magnetosphere*, 2007, Cambridge University Press, ISBN-0521035457.

Fischer, Daniel, *Mission Jupiter: The Spectacular Journey of the Galileo Spacecraft,* 2001, ISBN-0387987649.

McAnallu, John W. *Jupiter: and How to Observe It,*

ISBN-1852337508.

Sheehan, William; Hockey, Thomas *Jupiter*, 2018, ASIN-B07D3BBQ9Y.

Stakem, Patrick H.; Da Costa, Rodrigo Santos Valente; Rezende, Aryadne; Ravazzi, Andre "A Cubesat-based alternative for the Juno Mission to Jupiter," 2017, Presentation to Flight Software-15, JHU-APL.

https://solarsystem.nasa.gov/planets/Jupiter

http://nineplanets.org/Jupiter.html

Saturn

Meltzer, Michael *The Cassini-Huygens Visit to Saturn: An Historic Mission to the Ringed Planet,* 2015, ISBN-3319076078.

NASA, *The Saturn System Through The Eyes Of Cassini*, 2018, ISBN-1680922149.

Regius, Codex *Enceladus - Iceland of Space: The Cassini spacecraft over the moon of chilly geysers*, 2017, ASIN-B078R3J1VZ.

Regius, Codex *Titan: Pluto's big brother: The Cassini-Huygens spacecraft and the darkest moon of Saturn,* 2016, ASIN-B01N5JBUSK.

https://solarsystem.nasa.gov/planets/Saturn

http://nineplanets.org/saturn.html

Uranus

Glaser, Chaya *Uranus: Cold and Blue,* ISBN-1627245677.

https://solarsystem.nasa.gov/planets/uranus

http://nineplanets.org/uranus.html

Neptune

Miner, Ellis D.; Wessen, Randii R. *Neptune: The Planet, Rings, and Satellites* , 2002, Springer, ASIN-B01A0CEX14.

https://solarsystem.nasa.gov/planets/Neptune

http://nineplanets.org/neptune.html

Resources

- https://www.mars-one.com/
- WWW. Space-settlement-institute.org
- New Horizons Mission - https://www.nasa.gov/mission_pages/newhorizons/overview/index.html
- www.planetary.org
- https://nssdc.gsfc.nasa.gov/planetary/planetfact.html
- Wikipedia, various.
- New Horizons Mission - https://www.nasa.gov/mission_pages/newhorizons/overview/index.html
- Visions and Voyages for Planetary Science in the Decade 2013 – 2022 (2011). avail: https://www.nap.edu/catalog/13117/vision-and-voyages-for-planetary-science-in-the-decade-2013-2022.
- JPL Small-body Database. Avail:https://ssd.jpl.nasa.gov/sbdb.cgi
- https://pds-rings.seti.org/jupiter/
- http://astronomy.swin.edu.au/cosmos/C/Centaurs
- https://svs.gsfc.nasa.gov/
- https://voyager.jpl.nasa.gov/mission/science/
- http://www.iflscience.com/space/how-saturns-shepherd-moons-herd-its-rings/
- https://www.space.com/28640-living-on-ceres-asteroid-belt.html
- https://www.centauri-

dreams.org/2010/06/03/manned-missions-to-the-outer-system/
- Landis, Geoffrey (2011)"Terraforming Venus: A Challenging Project for Future Colonization" (PDF).Doi:10.2514/6.2011-7215, AIAA-2011-7215, AIAA Space 2011 Conference & Exposition, Long Beach CA, Sept. 26–29, 2011.
- http://www.redcolony.com
- Averner, M. M. & MacElroy, R. D., "On the Habitability of Mars: An Approach to Planetary Ecosynthesis.," 1976, NASA SP-414. Avail: https://ntrs.nasa.gov/search.jsp?R=19770005775.
- NASA's Advanced Planning and Integration Office, ISRU capability roadmap. https://ntrs.nasa.gov/archive/nasa/casi.ntrs.nasa.gov/20050204002.pdf
- https://www.nasa.gov/centers/ames/research/technology-onepagers/in-situ_resource_Utiliza14.html
- wikipedia, various.

If you enjoyed this book, you might also be interested in some of these.

Stakem, Patrick H. *Floating Point Computation*, 2013, PRRB Publishing, ISBN-152021619X.

Stakem, Patrick H. *Architecture of Massively Parallel Microprocessor Systems*, 2011, PRRB Publishing, ISBN-1520250061.

Stakem, Patrick H. *Multicore Computer Architecture,* 2014, PRRB Publishing, ISBN-1520241372.

Stakem, Patrick H. *Personal Robots*, 2014, PRRB Publishing, ISBN-1520216254.

Stakem, Patrick H. *RISC Microprocessors, History and Overview,* 2013, PRRB Publishing, ISBN-1520216289.

Stakem, Patrick H. *Robots and Telerobots in Space Application*s, 2011, PRRB Publishing, ISBN-1520210361.

Stakem, Patrick H. *The Saturn Rocket and the Pegasus Missions, 1965,* 2013, PRRB Publishing, ISBN-1520209916.

Stakem, Patrick H. *Visiting the NASA Centers, and Locations of Historic Rockets & Spacecraft,* 2017, PRRB

Publishing, ISBN-1549651205.

Stakem, Patrick H. *Microprocessors in Space*, 2011, PRRB Publishing, ISBN-1520216343.

Stakem, Patrick H. Computer *Virtualization and the Cloud*, 2013, PRRB Publishing, ISBN-152021636X.

Stakem, Patrick H. *What's the Worst That Could Happen? Bad Assumptions, Ignorance, Failures and Screw-ups in Engineering Projects*, 2014, PRRB Publishing, ISBN-1520207166.

Stakem, Patrick H. *Computer Architecture & Programming of the Intel x86 Family*, 2013, PRRB Publishing, ISBN-1520263724.

Stakem, Patrick H. *The Hardware and Software Architecture of the Transputer*, 2011, PRRB Publishing, ISBN-152020681X.

Stakem, Patrick H. *Mainframes, Computing on Big Iron*, 2015, PRRB Publishing, ISBN- 1520216459.

Stakem, Patrick H. *Spacecraft Control Centers*, 2015, PRRB Publishing, ISBN-1520200617.

Stakem, Patrick H. *Embedded in Space*, 2015, PRRB Publishing, ISBN-1520215916.

Stakem, Patrick H. *A Practitioner's Guide to RISC Microprocessor Architecture*, Wiley-Interscience, 1996, ISBN-0471130184.

Stakem, Patrick H. *Cubesat Engineering*, PRRB Publishing, 2017, ISBN-1520754019.

Stakem, Patrick H. *Cubesat Operations*, PRRB Publishing, 2017, ISBN-152076717X.

Stakem, Patrick H. *Interplanetary Cubesats*, PRRB Publishing, 2017, ISBN-1520766173 .

Stakem, Patrick H. Cubesat Constellations, Clusters, and Swarms, Stakem, PRRB Publishing, 2017, ISBN-1520767544.

Stakem, Patrick H. *Graphics Processing Units, an overview*, 2017, PRRB Publishing, ISBN-1520879695.

Stakem, Patrick H. *Intel Embedded and the Arduino-101, 2017,* PRRB Publishing, ISBN-1520879296.

Stakem, Patrick H. *Orbital Debris, the problem and the mitigation*, 2018, PRRB Publishing, ISBN-*1980466483*.

Stakem, Patrick H. *Manufacturing in Space*, 2018, PRRB Publishing, ISBN-1977076041.

Stakem, Patrick H. *NASA's Ships and Planes*, 2018,

PRRB Publishing, ISBN-1977076823.

Stakem, Patrick H. *Space Tourism*, 2018, PRRB Publishing, ISBN-1977073506.

Stakem, Patrick H. *STEM – Data Storage and Communications*, 2018, PRRB Publishing, ISBN-1977073115.

Stakem, Patrick H. *In-Space Robotic Repair and Servicing*, 2018, PRRB Publishing, ISBN-1980478236.

Stakem, Patrick H. *Introducing Weather in the pre-K to 12 Curricula, A Resource Guide for Educators*, 2017, PRRB Publishing, ISBN-1980638241.

Stakem, Patrick H. *Introducing Astronomy in the pre-K to 12 Curricula, A Resource Guide for Educators*, 2017, PRRB Publishing, ISBN-198104065X.
Also available in a Brazilian Portuguese edition, ISBN-1983106127.

Stakem, Patrick H. *Deep Space Gateways, the Moon and Beyond*, 2017, PRRB Publishing, ISBN-1973465701.

Stakem, Patrick H. *Exploration of the Gas Giants, Space Missions to Jupiter, Saturn, Uranus, and Neptune*, PRRB Publishing, 2018, ISBN-9781717814500.

Stakem, Patrick H. *Crewed Spacecraft*, 2017, PRRB

Publishing, ISBN-1549992406.

Stakem, Patrick H. *Rocketplanes to Space*, 2017, PRRB Publishing, ISBN-1549992589.

Stakem, Patrick H. *Crewed Space Stations,* 2017, PRRB Publishing, ISBN-1549992228.

Stakem, Patrick H. *Enviro-bots for STEM: Using Robotics in the pre-K to 12 Curricula, A Resource Guide for Educators,* 2017, PRRB Publishing, ISBN-1549656619.

Stakem, Patrick H. *STEM-Sat, Using Cubesats in the pre-K to 12 Curricula, A Resource Guide for Educators*, 2017, ISBN-1549656376.

Stakem, Patrick H. *Lunar Orbital Platform-Gateway*, 2018, PRRB Publishing, ISBN-1980498628.

Stakem, Patrick H. *Embedded GPU's*, 2018, PRRB Publishing, ISBN- 1980476497.

Stakem, Patrick H. *Mobile Cloud Robotics*, 2018, PRRB Publishing, ISBN- 1980488088.

Stakem, Patrick H. *Extreme Environment Embedded Systems,* 2017, PRRB Publishing, ISBN-1520215967.

Stakem, Patrick H. *What's the Worst, Volume-2*, 2018,

ISBN-1981005579.

Stakem, Patrick H., *Spaceports*, 2018, ISBN-1981022287.

Stakem, Patrick H., *Space Launch Vehicles*, 2018, ISBN-1983071773.

Stakem, Patrick H. *Mars*, 2018, ISBN-1983116902.

Stakem, Patrick H. *X-86, 40th Anniversary ed*, 2018, ISBN-1983189405.

Stakem, Patrick H. *Lunar Orbital Platform-Gateway*, 2018, PRRB Publishing, ISBN-1980498628.

Stakem, Patrick H. *Space Weather*, 2018, ISBN-1723904023.

Stakem, Patrick H. *STEM-Engineering Process*, 2017, ISBN-1983196517.

Stakem, Patrick H. *Space Telescopes,* 2018, PRRB Publishing, ISBN-1728728568.

Stakem, Patrick H. *Exoplanets*, 2018, PRRB Publishing, ISBN-9781731385055.

Stakem, Patrick H. *Planetary Defense*, 2018, PRRB Publishing, ISBN-9781731001207.

Patrick H. Stakem *Exploration of the Asteroid Belt*, 2018, PRRB Publishing, ISBN-1731049846.

Patrick H. Stakem *Terraforming*, 2018, PRRB Publishing, ISBN-1790308100.

Patrick H. Stakem, *Martian Railroad,* 2019, PRRB Publishing, ISBN-1794488243.

Patrick H. Stakem, *Exoplanets,* 2019, PRRB Publishing, ISBN-1731385056.

Patrick H. Stakem, *Exploiting the Moon,* 2019, PRRB Publishing, ISBN-1091057850.

Patrick H. Stakem, *RISC-V, an Open Source Solution for Space Flight Computers,* 2019, PRRB Publishing, ISBN-1796434388.

Patrick H. Stakem, *Arm in Space*, 2019, PRRB Publishing, ISBN-9781099789137.

Patrick H. Stakem, *Extraterrestrial Life*, 2019, PRRB Publishing, ISBN-978-1072072188.

Patrick H. Stakem, *Space Command*, 2019, PRRB Publishing, ISBN-978-1693005398.

CubeRovers, A Synergy of Technologys, 2020, PRRB

Publishing, ISBN-979-8651773138.

Robotic Exploration of the Icy moons of the Gas Giants. 2020, PRRB Publishing, ISBN- 979-8621431006

Hacking Cubesats, 2020, PRRB Publishing, ISBN-979-8623458964.

History & Future of Cubesats, PRRB Publishing, ISBN-979-8649179386.

Hacking Cubesats, Cybersecurity in Space, 2020, PRRB Publishing, ISBN-979-8623458964.

Powerships, Powerbarges, Floating Wind Farms: electricity when and where you need it, 2021, PRRB Publishing, ISBN-979-8716199477.

Hospital Ships, Trains, and Aircraft, 2020, PRRB Publishing, ISBN-979-8642944349.

2020/2021 Releases

CubeRovers, a Synergy of Technologys, 2020, ISBN-979-8651773138

Exploration of Lunar & Martian Lava Tubes by Cube-X, ISBN-979-8621435325.

Robotic Exploration of the Icy moons of the Gas Giants,

ISBN- 979-8621431006.

History & Future of Cubesats, ISBN-978-1986536356.

Robotic Exploration of the Icy Moons of the Ice Giants, by Swarms of Cubesats, ISBN-979-8621431006.

Swarm Robotics, ISBN-979-8534505948.

Introduction to Electric Power Systems, ISBN-979-8519208727.

Centros de Control: Operaciones en Satélites del Estándar CubeSat (Spanish Edition), 2021, ISBN-979-8510113068.

Exploration of Venus, 2022, ISBN-979-8484416110.

Patrick H. Stakem, *The Search for Extraterrestial Life,* 2019, PRRB Publishing, ISBN-1072072181.

The Artemis Missions, Return to the Moon, and on to Mars, 2021, ISBN-979-8490532361.

James Webb Space Telescope. A New Era in Astronomy, 2021, ISBN-979-8773857969.

www.ingramcontent.com/pod-product-compliance
Lightning Source LLC
Chambersburg PA
CBHW020622220526
45463CB00006B/2650